# Grouping

by Jennifer Boothroyd

Lerner Publications Company · Minneapolis

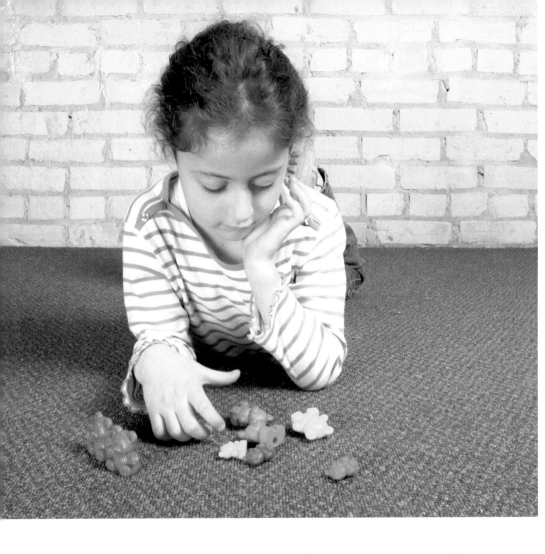

How can I make groups?

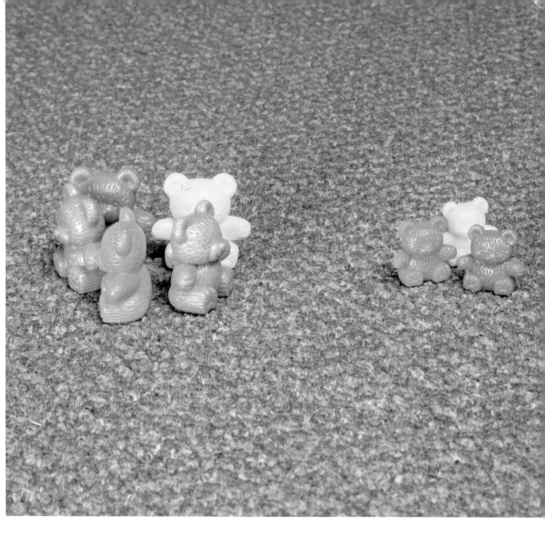

I make groups by size.

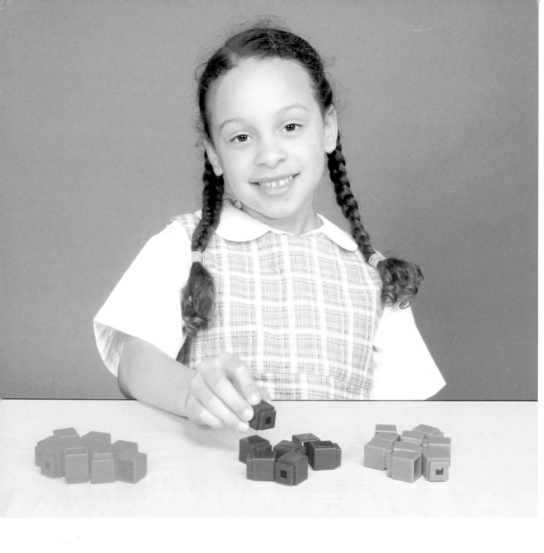

How can I make groups?

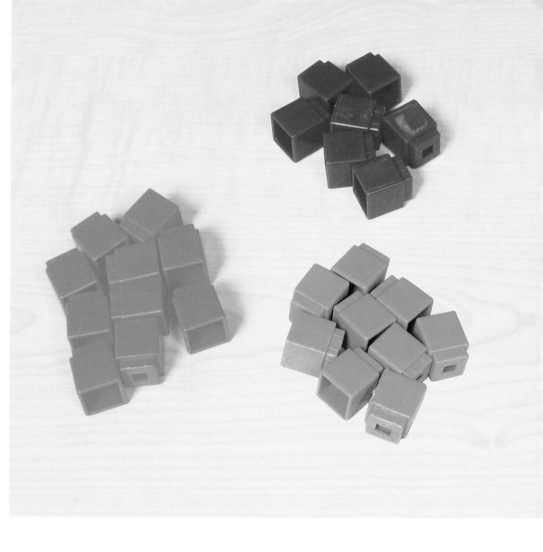

I make groups by color.

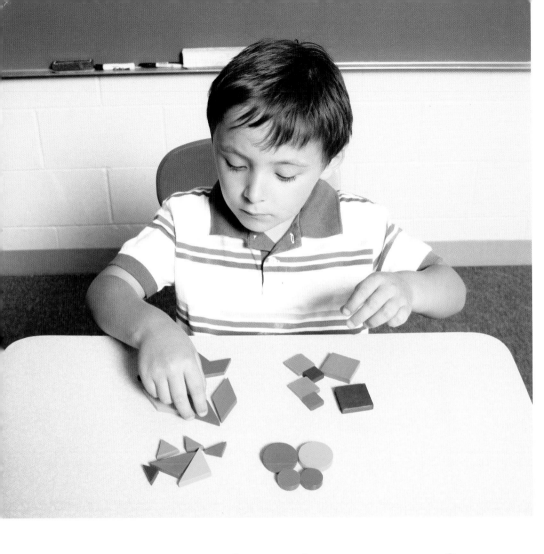

How can I make groups?

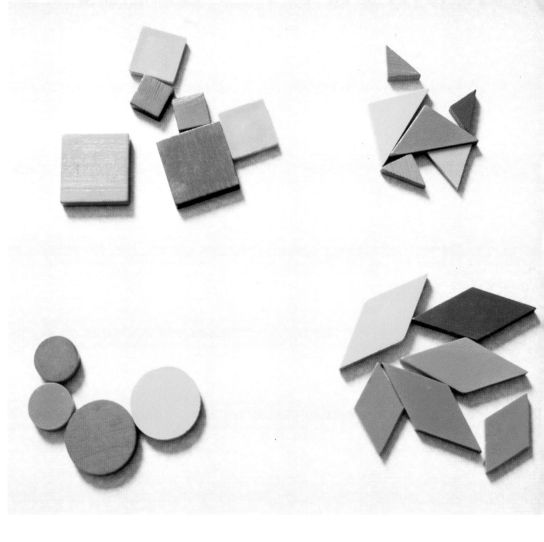

I make groups by shape.

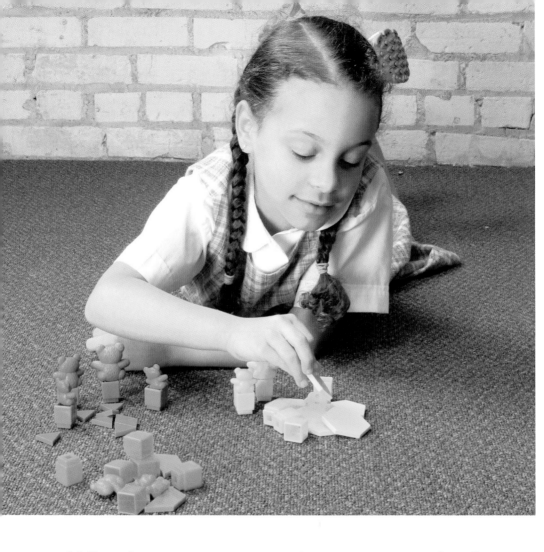

What groups can you make?